BRUNEL
IN CORNWALL

JOHN CHRISTOPHER

AMBERLEY

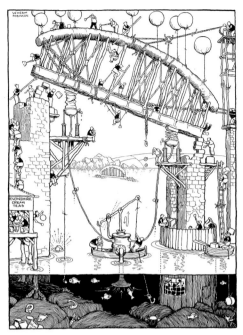

The construction of Saltash Bridge, depicted by
Mr. W. Heath Robinson

Left: W. Heath Robinson's version of how they
raised the spans at Saltash. Taken from the 1935
GWR publication, *Track Topics.*

About this book

Hopefully this book will encourage you to delve
a little deeper into IKB's works in Cornwall, but
please note that some sites may be on private
property and access might be restricted for
reasons of safety and security.

First published 2014

Amberley Publishing
The Hill, Stroud
Gloucestershire, GL5 4EP

www.amberley-books.com

Copyright © John Christopher, 2014

The right of John Christopher to be identified as the
Author of this work has been asserted in accordance
with the Copyrights, Designs and Patents Act 1988.

ISBN 978 1 4456 1859 3 (print)
ISBN 978 1 4456 1878 4 (ebook)

British Library Cataloguing in Publication Data.
A catalogue record for this book is available from
the British Library.

Typeset in 9.5pt on 12pt Celeste.
Typesetting by Amberley Publishing.
Printed in the UK.

Brunel's Cornwall

'Kernow a'gas dynnergh.' Welcome to Cornwall, proclaims the sign on the platform at Saltash station beneath an image of the Royal Albert Bridge. Cornwall is at the south-western extreme of Brunel's kingdom, an area defined by the spidery web of his railway lines that spread from London down to Penzance, across South Wales to Pembrokeshire, and northwards into the Midlands and also down towards the south coast. This was almost exclusively broad gauge territory and consequently Brunel ruled it as supreme engineer until his death in 1859, with his influence continuing for many decades afterwards. In fact it would be fair to say that it continues to this day – just ask the people of South Devon and Cornwall who were cut off by the collapse of the sea wall at Dawlish in early 2014. The coming of the railway to Cornwall shaped the county, physically through the bridges, viaducts and cuttings, which left an indelible mark on the landscape, and also through the development of its industries and the creation of a new one, tourism. When George Bradshaw's famous *Descriptive Railway Hand-book of Great Britain and Ireland*, now universally known simply as Bradshaw's Guide, was published in 1863, it had described Cornwall in less than glowing terms:

> Cornwall, from its soil, appearance, and climate, is one of the least inviting of the English counties. A ridge of bare and rugged hills, intermixed with bleak moors, runs through the midst of its whole length, and exhibits the appearance of a dreary waste.

Below: The sign at Saltash station welcomes travellers in two languages. Sadly, the plans to restore the main station building as a visitor centre appear to have been abandoned.

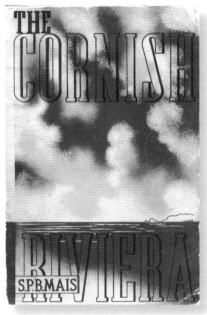

It was the Great Western Railway that coined the term 'Cornish Riviera' as a marketing ploy to convince visitors of the duchy's climatic charms. It was to feature on a wide range of promotional items, as shown here.

Left top: An early advertising postcard, c. 1905. *(CMcC)*

Left bottom: And another early advertisement, this time comparing Cornwall with Italy no less.

Above: A later edition of the GWR guidebook by A. P. B. Mais, this one published in 1934.

Opposite page: Enamel sign, showing the company's routes within Cornwall, displayed at the Steam museum in Swindon, plus two posters.

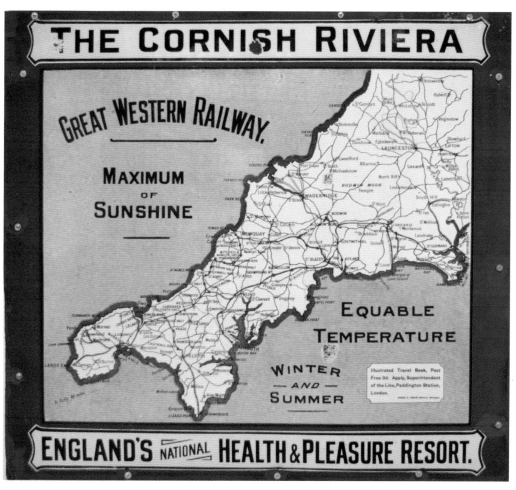

THE CORNISH RIVIERA

GREAT WESTERN RAILWAY.

MAXIMUM OF SUNSHINE

EQUABLE TEMPERATURE

WINTER — AND — SUMMER

ENGLAND'S NATIONAL HEALTH & PLEASURE RESORT.

GO GREAT WESTERN TO CORNWALL

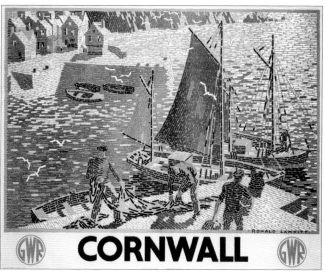

CORNWALL

How times have changed. Certainly, when the guide was published the railway into Cornwall had only been operational for barely four years following the completion of the Royal Albert Bridge over the River Tamar. The initial purpose of the county's railways was to serve its industries, the tin, copper and lead mines in the west, and to provide a connection between Falmouth and London. For many years Falmouth was handling nearly all of the packet trade. These were the despatches coming into the country from various parts of Britain's overseas territories, and from Falmouth they had to be taken on to London by road. Unfortunately Falmouth had a serious rival for this trade in the form of Southampton, and once a rail link had been established between Southampton and London, in 1840, the government moved the bulk of the packets trade to Southampton. This took away much of the expected income for the proposed Cornish rail-link and caused a delay in its completion. But thanks to Brunel and to Bradshaw a new user of the railways was to emerge and Cornwall was to become the flagship tourist destination of the Great Western Railway. On company posters it was compared with Italy in terms of its shape, climate, and natural features, beneath a banner proclaiming 'See Your Own Country First'. Another described it as 'Britain's National Health and Leisure Resort'. It was a magnificent piece of creative marketing, branding the coastal resorts as the 'Cornish Riviera'. The GWR published several of its own guidebooks in the 1920s and 1930s and this is how the introduction to eponymous *Cornish Riviera* guide reinforced the message:

> Cornwall is recognised as an area where visitors may be reasonably assured that they will escape the rigours of the winter ... It is because its climate, the all-important factor in a holiday, is the most equable in the world that we claim the right to describe the Duchy as the Cornish Riviera.

While the railways form the backbone of this account of Brunel's works in Cornwall, the county is no less interesting for its other Brunellian connections. One of the advantages of this series of books examining Brunel's legacy in individual cities or regions – and this is the fourth volume, with the previous ones covering Gloucestershire, Bristol and London – is that it provides the opportunity to look at the more local connections overlooked by the more weighty accounts of his life and work. Each area has thrown up its own surprises, and Cornwall is no exception. As we will see, the Cornish link runs through his engineering career like a seam, starting miles away under the murky water of the Thames.

The Cornish tunnellers

It was Marc Brunel, Isambard's father, who in 1825 brought Cornish miners to London to work on the world's first subaquatic tunnel. The plan was to drive a tunnel underneath the Thames to provide access to the Surrey Commercial Docks on the south side of the river as an alternative to the increasingly congested road traffic. At that time London Bridge was the only crossing point on that side of the city. There had been previous attempts at tunnelling beneath rivers, but all had been unsuccessful. In 1805 Robert Vazie, a Cornish tin mining engineer with experience of driving mineshafts under the sea, constructed a small pilot tunnel or 'driftway', just 3 feet by 3 feet in cross section, under the Thames, but the works were overwhelmed by water. Another celebrated Cornishman, Richard Trevithick, took over and his miners got within 200 feet of the opposite shore before their tunnel flooded and the project was abandoned. The major breakthrough in tunnelling methods came with Marc Brunel's design for a tunnelling shield. The 'Great Shield', as it was known, consisted of twelve vertical iron frames, each 3 feet wide and divided into three cells, one above the other, having a miner working in each. At the front of the shield was a series of horizontal boards which could be removed individually to expose only a small area to be excavated at any one time, thus reducing the risk of a cave-in. These boards were then moved forward and, once they had all been repositioned, the frame would be shunted forward by jacks pressing against the tunnel brickwork built to fill the gap behind the shield as it progressed.

The first task in constructing the Thames Tunnel was to create an access shaft at Rotherhithe on the southern side of the river. Marc achieved this by building a huge 3-foot-thick brick cylinder, 50 feet in diameter and 42 feet high, which was braced with vertical metal rods and had iron rims top and bottom. Acting like a massive pastry cutter this cylinder sank under its own weight as the ground was excavated beneath it. Once the shaft was completed the shield was put in place and the work on boring the horizontal tunnel commenced. Conditions for the miners, mainly recruited from the Cornish tin mines, and also from the Durham coalfields, were truly appalling, with the filthy river water constantly seeping in through the gaps. The men retched in the foul air and many were struck down by a condition known as 'tunnel sickness', which sometimes caused blindness. Inevitably the terrible conditions took their toll and by April 1826 Marc's resident engineer, William Armstrong, had resigned. Marc's health was also suffering by this time and there was one obvious candidate to take over the job.

Isambard Kingdom Brunel became resident engineer on Europe's most demanding engineering project when he was barely twenty years of age. He immediately threw himself into the task, spending more than twenty hours a day working within the tunnel, taking only catnaps to keep going, often for days on end. From the start IKB was very much a hands-on engineer and it was not uncommon for him to work shoulder to shoulder with the

Right: Contemporary illustration of a tunneller, together with a profile portrait of Sir Marc Brunel, published at the time of the tunnel's opening. Brunel senior received a knighthood from Queen Victoria, an honour not bestowed upon his more famous son.

THAMES TUNNEL, OPEN'D 25 MARCH 1843

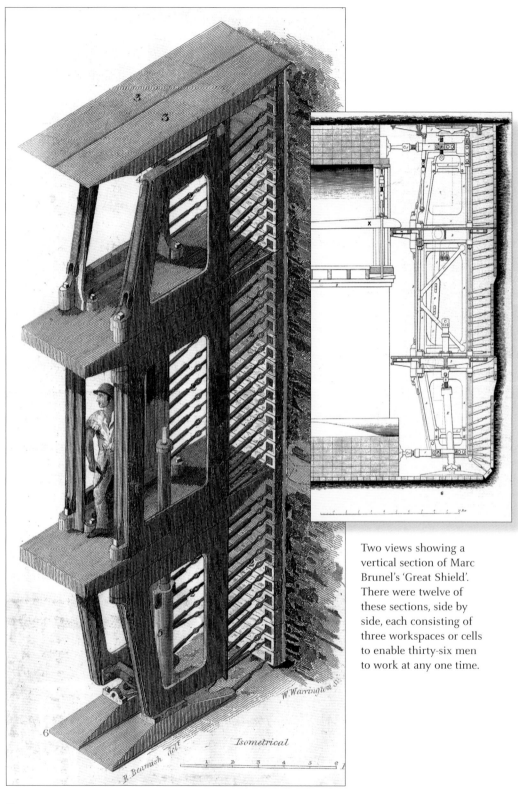

Two views showing a
vertical section of Marc
Brunel's 'Great Shield'.
There were twelve of
these sections, side by
side, each consisting of
three workspaces or cells
to enable thirty-six men
to work at any one time.

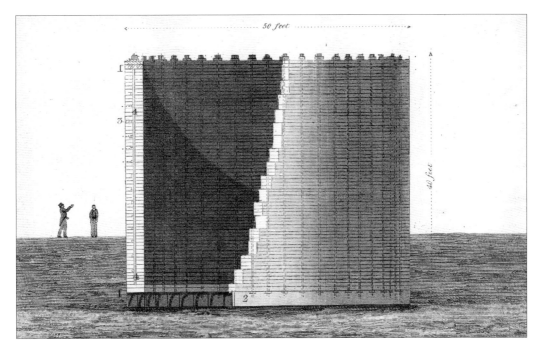

Above: On the Rotherhithe side of the river this massive brick tube acted like a pastry cutter to create an access shaft as the soil was excavated. A similar method was also used on the Wapping side, and that tube remains in place as the stairwell down to the platform level.

miners. By the beginning of 1827 progress was good and they had driven the tunnel 540 feet under the Thames, almost halfway to Wapping on the north bank. But both Brunels became increasingly concerned about the risk of breakthrough as the miners encountered more and more gravel instead of the seam of firm London clay they had anticipated. Furthermore, the gravel was often littered with river debris, such as discarded clay pipes, indicating that the riverbed was perilously close to the workings. Marc estimated that the shield was holding up to 600 tons of water at bay and his concerns were exasperated by the presence of a flow of shilling-sightseers admitted into the works by the company directors. It was only by sheer good fortune that there were no members of the public present when the river broke through on the evening of 18 May 1827. IKB had been working a little way back from the shield when he looked up to see the panic-stricken workmen running for their lives. The surge of water extinguished the lights and in the blackness and deafening din he did his best to get everyone to the access shaft as quickly as possible, but as they neared the top of the shaft some of the steps were swept away. Suddenly a cry for help was heard, and tying a rope around his waist Isambard slid down one of the iron ties of the shaft to rescue an old engineman named Tillett.

Incredibly, there were no casualties that day. To investigate the extent of the damage, IKB borrowed a diving bell from the West India Dock Company and descended beneath the cold waters to the riverbed where he was able to place one foot on the No. 12 frame of the shield and the other on the exposed brickwork. Over the ensuing months the hole was plugged with a combination of iron rods and 150 tons of clay-packed bags, and as the pumps did their work

Above: Photographed in 2005, the original twin horseshoe entrances to the Thames Tunnel are still visible from the Wapping platform. Note how the tunnel dips towards the middle of the river, a feature not shown in comtemporary illustrations. It was built for the transportation of cargo from the new docks on the south side of the river, but access ramps for horse-drawn carts were never built, and after a period as a pedestrian tunnel it was sold to the East London Railway in 1865.

Left: On the Rotherhithe side, this building originally housed one of the steam engines that pumped water out of the tunnel workings. It is now part of the excellent Brunel Museum.

and the water slowly receded until he was able to reach the shield by boat inside the tunnel and by crawling on the bank of debris.

By the end of the year the workings had been cleared in time for a celebratory Christmas banquet held in the tunnel itself. This was a carefully contrived publicity stunt intended to restore confidence in the project. Tunnelling was resumed, but on 12 January 1828 the tunnel flooded again. This time the torrent was so violent that the rush of water carried Isambard, and his companions, up to the lip of the shaft where he was plucked to safety by the others. Despite sustaining serious injuries, both to his leg and internally, he called for the diving bell once more but this time he had to direct operations from a mattress on the barge. The damage to the tunnel was far greater than with the first deluge and with funds dwindling the directors ordered the workings to be bricked up, much to the derision of the newspaper hacks who summed up the whole business as 'the great bore'. With the tunnel closed, Isambard was sent to Brighton to recuperate and afterwards he went on to Bristol to follow his own career. It was in an uncharacteristically despondent mood that he made this entry in his journal:

> Tunnel is now, I think, dead … This is the first time I have felt able to cry at least for these ten years. Some further attempts may be made – but – it will never be finished now in my father's lifetime I fear.

He never worked on the tunnel again, but his predictions about it being dead proved to be wrong. The work was resumed and thanks in no small measure to the Cornish miners the Thames Tunnel was completed and officially opened to the public on 25 March 1843 when it was heralded as the 'eighth wonder of the world'. In 1865 it was sold to the East London Railway and five years later reopened as the first subaquatic railway tunnel in the world. The stations at Wapping and Rotherhithe were part of the East London Line on the London Underground until April 2010 and following completion of the Phase 1 extension of that line the tunnel is now part of the new London Overground network, albeit beneath the Thames.

Explore the Thames Tunnel

• Wapping station incorporates the tunnel shaft with access via a stairway to the platforms. Note that these are probably the narrowest platforms to be found anywhere in London. The horseshoe-shaped entrance arches are clearly visible.

• Rotherhithe station is built a little further back from the tunnel and rail access is via a later extension, so you can't see the tunnel itself.

• The excellent Brunel Museum is located around the corner from the Rotherhithe station (turn left on leaving the station) and is housed in the original pumping engine house beside the tunnel shaft. The shaft is directly above the railway lines, but a new floor has been fitted and there are plans to incorporate this truncated cylindrical space within the museum. Various events are held at the museum throughout the year including some tours of the tunnel itself. The museum is open seven days a week – check the website for details, www.brunel-museum.org.uk.

• Of course you can always take the train through the tunnel.

Bridging the Tamar

Brunel's involvement with the railways of Cornwall had begun in February 1845 when he appeared before a parliamentary committee examining a Bill submitted by the Cornwall Railway. He was already engineer to the South Devon Railway, and the plan was to extend the existing coastal line westwards from Devon with the backing of the so-called Associated Companies – mainly the Cornish railway, the GWR and other broad gauge companies. The primary objective was to take it from Plymouth as far as Falmouth, but the proposed line had already missed the heights of 'railway mania' in the early 1840s and once Falmouth had lost out on the packet trade there was a hiatus of several years before interest in a railway revived. (See the section on the Falmouth Branch.)

By this time the Bristol & Exeter Railway had reached Exeter (in May 1844), and the South Devon Railway was working on the connection with Plymouth (opened in April 1849). The prospect of linking up with a Cornish line to create a through route all the way from London was still an attractive one, and the Cornwall Railway and the GWR agreed that a southern route through Cornwall was the most suitable. The main obstacle in the way was the Hamoaze, the wide body of water at the mouth of the River Tamar. One plan was to load the trains onto special ferries, although this would have proved a very awkward procedure as the fast-flowing waters at this point have a rise and fall of 18 feet. Even so, Brunel gave his tacit approval of the scheme, as he said to the parliamentary committee, 'I am prepared to say that I consider there is no difficulty in doing it.' However, as Brunel must have anticipated, the ferry plan fell through and the directors of the railway went away to reconsider their options. Spurred on by a rival plan to take the line further inland via Oakhampton, Brunel was appointed as engineer to the Cornwall Railway in August 1845. He immediately started a new survey of the route, incorporating a bridge over the Tamar at Saltash, and by the time he reappeared before another parliamentary committee the rival inland scheme had fizzled out. With Brunel on board the Cornwall Railway Act received royal assent in August 1846.

The exact form of the bridge over the 1,000-foot-wide stretch of the river had yet to be determined. Brunel had considered a massive single span, possibly built of timber, in order to meet Admiralty stipulations regarding clearance for its tall-masted ships. (Timber would come to feature prominently in this story.) However, within a year of the Act being passed, the first flush of 'railway mania' had run its course and there was a slump in the value of railway company shares and, consequently, in the interest of investors. The Cornish railway was put on hold temporarily and this allowed Brunel to pursue his other projects, while also giving him a breathing space in which to develop his ideas for the Tamar crossing. Crucially, this time was well spent in building the tubular iron bridge for the railway over the River Wye at Chepstow and this would serve as a prototype for the Tamar bridge.

When Brunel returned to the Cornish project, he settled upon a tubular bridge design featuring two identical spans of 465 feet, although this was reduced to 455 feet later on, with a single mid-river pier. On dry land on either side there would be a main pier accessed by a series of girder spans carried on several pairs of smaller piers. The biggest challenge

Before Saltash, Brunel had experiemented with tubular wrought-iron construction on two previous bridges.

Top: The railway bridge at Chepstow takes the line over the River Wye into South Wales. Opened for trains in 1852, it was dismantled in 1962.

Middle: Detail of Brunel's swing bridge, built to cross his South Entrance Lock into Bristol's Floating Harbour. It is still there, along with a later duplicate bridge.

Bottom: Robert Stephenson's tubular bridge over the Menai Straits was completed in 1850, pre-dating the Chepstow bridge by two years. Unlike Brunel's design, the railway tracks are contained within the tube itself. *(CMcC)*

Opposite page: The Royal Albert Bridge seen from Saltash station, 2005.

was the construction of a central pier mid-river and, in order to investigate the nature of the riverbed, in 1848 he had a wrought-iron tube constructed, 85 feet long and 6 feet in diameter. Acting like a cofferdam, this tube was lowered into the water until one end rested on the riverbed, and it was pumped out to enable the drilling of a number of trial borings. Fortunately these revealed a bed of hard rock beneath a thick layer of mud and slime. Armed with this information Brunel was able to further his designs, but once again a lack of funding meant that matters were put on hold for three years. In 1851 he attempted to resuscitate the Cornwall Railway project by proposing that savings could be made by reducing the line to a single track, including that on the Saltash bridge. This was approved in 1852, by which time Brunel's designs for the bridge had been finalised in the form we see today.

The Royal Albert Bridge at Saltash has been described as a 'masterpiece of complexity' and it is easy to see why. As with Chepstow's bridge, it is a closed structure incorporating three engineering forms: the compression tubular arch, the tension chains of a suspension bridge, and a beam deck. In the photographs it is easy to make out the suspension chains, which mirror the twin humps of the tubular girders. Look again at the towers, and you will see how closely they resemble those of the Clifton Suspension Bridge, and the connection goes further than that as some of the chains came from Bristol when that bridge was left unfinished. However, the big difference is that at Bristol the chains pass over the towers and are anchored into the rock on either side creating an 'open' structure where the forces are equalised. With the double span at Saltash there is nowhere for the chains to go and so their

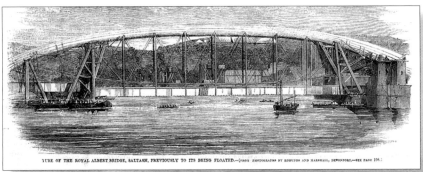

TUBE OF THE ROYAL ALBERT BRIDGE, SALTASH, PREVIOUSLY TO ITS BEING FLOATED.—(FROM PHOTOGRAPHS BY EDMUNDS AND MARSHALL, DEVONPORT.—SEE PAGE 196.)

The two main spans at Saltash were floated into position on pontoons and then raised up using hydraulic jacks and by building up the corresponding masonry pier 3 feet at a time. The top image shows the newly positioned span on the Devon side.

tension forces, in effect pulling the towers towards each other, are countered by the tubular arches at the top. It is a bowstring design, similar in some respects to his bridge at Windsor, but with an extra twist in the design in the form of the chains. Furthermore, the arched girder is much more substantial, and to further stiffen the structure Brunel incorporated vertical hangers or struts running from the deck to the arches, with additional diagonals for longitudinal stability.

The part of the central pier visible above the waterline consists of four octagonal iron columns 96 feet high, but beneath them is a masonry column descending a further 96 feet to the riverbed. To construct this, Brunel built another cylinder, a much bigger one at 35 feet wide. In effect this functioned like a diving bell in which the stonemasons could work. To keep the river water from seeping in, the air within the cylinder was kept under pressure by steam-driven pumps, and this made it possible for them to work 70 feet beneath the water, although they frequently suffered the severe headaches and muscular pains we now associate with the 'bends'. The central pier was completed by 1856 and the great cylinder was unbolted, split into two and removed. In comparison the construction of the two main side piers, the portal piers, and the seventeen narrower piers on the approach spans – eight for Devon and nine on the Cornwall side – was a far simpler matter and would be carried out in conjunction with raising the two main spans. The portal towers at each end of the bridge have a masonry and brick lining within an iron shell, while the equivalent portal on the central pier is hollow and made of wrought iron.

While work progressed on the various piers, the first of the two trusses, or girders, was taking shape on the Devon shore. The curving upper girders were oval in cross section being 12 feet 3 inches high and 16 feet 9 inches across. They are hollow and it is possible for a workman to pass through their entire length. Their oval cross section served the dual purpose of reducing sideways wind resistance and also permitting the deck struts to hang vertically. Each truss weighed 1,000 tons and was to be positioned at the bottom of its piers by floating it into position on huge pontoons, which could be raised or lowered in the water by pumping water in or out of them. The day for moving the first truss to the Cornwall side of the bridge was set for 1 September 1857. Brunel had made meticulous plans for the operation and he took command from a platform mounted high up on the truss, accompanied by a team of flag-waving signallers to convey his instructions to the various teams of workmen. Thousands of onlookers, some estimates suggest 300,000, crammed the shoreline and hills to witness this wonder, but they had been warned that absolute silence was demanded by the great engineer. As the signal flags fluttered out their commands, one eyewitness recorded the event:

Not a voice was heard, not a direction spoken; a few flags waved, a few boards with numbers on them were exhibited, and, by some mysterious agency, the tube and rail borne on pontoons travelled to their resting-place, and with such quietude as marked the building of Solomon's temple. With the impressive silence which is the highest evidence of power, it slid, as it were, into position without an incident, without any extraordinary effort, without a 'misfit' to the extent of the eighth of an inch. Equally without haste and without delay, just as the tide reached its limit, at 3 o'clock, the tube was fixed on the piers 30 ft above high water.

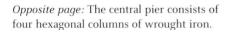

Opposite page: The central pier consists of four hexagonal columns of wrought iron.

This page, anti-clockwise from the top: Hand-coloured photograph, *c.* 1895. *(LoC)* Terence Cuneo's artwork for the 1963 Triang catalogue. Cover of the GWR's 1935 publication *Track Topics.*

Above left: This fine bust of Brunel is located at the bottom of Fore Street in Saltash. It is based on Baron Marochetti's statue on the Victoria Embankment, London. Alas the other statue at Saltash, the green man, is a poor portrayal of the great engineer. *Below:* The bridge photographed from the Devon shore, *c.* 1900, long before the arrival of the A38 road bridge spoilt the view.

This was Brunel's moment. He was the master magician, in complete charge as he reshaped the landscape to his bidding. When the band struck up 'See the Conquering Hero Comes', the spell of silence was broken and the vast crowd erupted into a cacophony of cheering.

The process of raising the truss to its final position was a far more protracted matter, with huge hydraulic jacks lifting it 3 feet at a time as the iron sections were added to the central pier and the masonry was put in place on the land pier. The mortar had to be given time to fully set between each lift and it wasn't until May the following year, 1858, that it was finally in position. By the time that the second span, the one for the Devon side, was ready to be moved, Brunel was in London struggling with the Great Eastern steamship. Accordingly his most trusted assistant, Robert Brereton, supervised the floating of the span and its positioning in July 1858, and it was under his guidance that the bridge was completed in early 1859. By this time Brunel's health was in decline as he was suffering from a kidney condition known at the time as Bright's disease. When Prince Albert travelled by GWR train down from Paddington to open the Royal Albert Bridge, as it was to be known, in May of 1859, Brunel was too ill to attend the ceremony. Shortly afterwards he saw the completed bridge for the first and last time. Reclining on a couch placed on an open trolley, he was drawn slowly across the bridge by one of Daniel Gooch's steam locomotives. It was a poignant moment. 'No flags flew, no bands played, no crowds cheered ...'

Brunel died in London just a few months later on 15 September 1859 and it was Brereton who supervised the continuing railway works within Cornwall. As a tribute to the great engineer the directors of the Cornwall Railway had the following epitaph added to the portals at each end of the bridge: 'I. K. BRUNEL ENGINEER 1859.'

Explore Saltash and the bridge

• The most conspicuous addition since Brunel's day is the A38 Tamar Road Bridge, a suspension bridge as it happens, which opened in 1961. Although it clutters the view from some angles, it does provide a superb platform from which to see the railway bridge. There is a parking area on the Devon side and a pathway on the bridge for pedestrians and cyclists, but note that cars going into Devon pay a toll.

• In Saltash head down to the shoreline to get the most spectacular view of the bridge and of the piers. While there you can't fail to notice the green man, arguably the worst Brunel statue to be found anywhere, and there is a plaque near the base of the land piers. The station building (rebuilt by the GWR in 1880) is boarded up, but the platforms provide a good view of the bridge and approach. Back up to street level there is the bust of Brunel based on the Marochetti statue on the Victoria Embankment. Set back slightly, it is located at the bottom of Fore Street as it turns towards the dual carriageway.

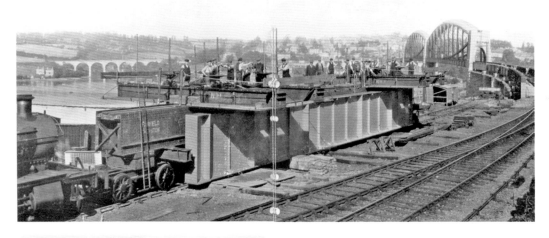

A fascinating group of images reproduced from the GWR's 1935 publication *Track Topics*, subtitled 'A Book of Railway Engineering for Boys of All Ages'.

The photographs on this page show the replacement of main girders on the fifteen approach or land spans in 1928:

'The actual job of replacing the land spans was carried out on Sundays when the traffic was comparatively light. The time of certain through trains was temporarily rearranged, and the local services at Saltash were run between Plymouth and St Budeaux stations whence passengers were conveyed by road motors to the ferry across the Tamar.'

Above: Interior of one of the main tubes.

Top right: Looking up from track level. Compare this with the modern photograph on page 25.

Below and bottom right: A workman walks on the top of one of the great humps. He has no safety harness and merely hooks one foot under a low-level rail.

Up close and
personal.

Opposite page:
The central
pier and portal
photographed
from the A38
Tamar Road
Bridge in 2005.

Right: Looking
up from the deck
level towards the
underside of one
of the spans. Note
the additional
cross-bracing. Part
of the chains can
be seen on the
lower left. It is
said that before
the arrival of the
road bridge, in
1961, the locals
would routinely
walk home across
the rail bridge
after a late night
out. *(Network Rail)*

Right:
One of the main
spans silhouetted
against sunlight
on the Hamoaze.

The Cornwall Railway

Work on the Cornwall Railway began in 1852, even though it would be another seven years before the Royal Albert bridge would be completed. The chosen route was particularly challenging in terms of its undulating terrain, and with money tight Brunel devised a series of designs for timber viaducts to cross the many valleys. Although a man of iron he was not afraid to work in other materials and had already built numerous timber viaducts and bridges on the Great Western Railway and on a number of other lines. He had also conducted a series of experiments to ascertain the strength of larger timbers and explored techniques for preserving the wood, in particular the Kyanising method that involved soaking pre-cut timbers in perchloride of Mercury. This led to far greater use of the material as the railway network spread across southern Wales and, later on, south-westwards into Devon and particularly in Cornwall. For the most part these timber viaducts on the Cornish Railway consisted of masonry piers 60 feet apart, centre to centre, rising up to within 35 feet of the decking. The piers were capped with iron plates from which timber struts radiated like an upturned hand, with further cross-bracing to support longitudinal beams beneath the deck. There were a number of variations to the timber structures, and these have been classified and grouped by type with further subdivisions based on specific structural details. For example a CLB4(A) type indicates a continuous laminated beam (CLB) fan viaduct with only two beams and two fans across the width of the viaduct. While these classifications are included here, it is beyond the scope of this book to reiterate the entire list and readers are referred to the two seminal works on the subject, *Brunel's Timber Bridges and Viaducts* by Brian Lewis, and *Brunel's Cornish Viaducts* by John Binding. (See the list on page 95.)

On the whole the design of the individual timber components was standardised, partly to facilitate easy replacement. These timber viaducts were never expected to last forever, and during their lifetime special maintenance teams became adept at replacing individual struts. The ever rising cost of repairing and maintaining the timber viaducts – exacerbated by a sharp rise in the cost of Baltic pine combined with concerns about their long-term safety and the need to double the track – meant that all of them were replaced, the majority by the 1880s and 1890s. In most cases new masonry viaducts were built alongside the old ones, but with some examples the original piers were retained by raising their height with brick or masonry to support iron girders. Either way, Brunel's Cornish viaducts have become an integral part of the landscape. As L. T. C. Rolt put it in his celebrated biography of Brunel:

> In the primeval, storm-bitten landscape of western Cornwall the tapering piers of local stone looked as much at home as the gaunt chimney stacks of the tin mines.

Opposite: Detail of the wooden structure on the Carvedras viaduct in Truro, with the timbers fanning out from the top of the masonry piers like an upturned hand.

Despite his reputation as a man of iron, Brunel made extensive use of timber construction on a number of his bridges, not just in Cornwall, and a selection is shown here. *Above:* The road bridge over Sonning Cutting on the GWR was one of his earliest. *Below:* The two-span skew bridge across the Avon at Bath had six laminated timber arches infilled with decorative ironwork.

Above: An engraving published in *The Engineer* in June 1892, just a month after the demise of the broad gauge. It illustrates two of the viaducts in South Devon, including, top right, the Monksmoor Bridge, which featured the 'King' truss commonly used on smaller bridges. The other images show the Ivybridge viaduct, before and during its replacement. *Below:* The longest of Brunel's timber structures was the viaduct at Landore, crossing the Swansea Valley in South Wales. At 1,760 feet long, it had a central span of 100 feet.

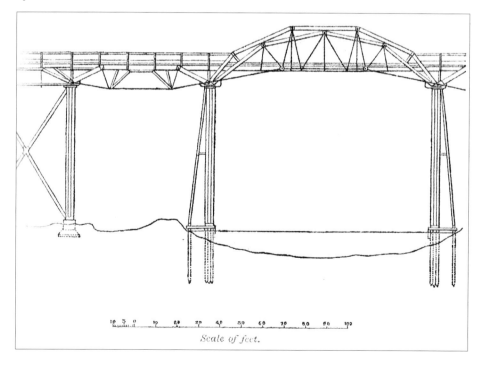

Scale of feet.

29

Photographs of the land spans on the Saltash side, separated in time by about 100 years. These impressive spans, with their straight double-legged piers, tend to be overshadowed by the bridge. *(CMcC/JC)*

Saltash to Liskeard

Having crossed the River Tamar, the line curves to the left, carried aloft above the roofs of Saltash on an elegant series of double-column piers of the type also seen in South Devon. Landfall brings it to the station, where a bilingual sign welcomes travellers to Cornwall.

Saltash station opened on 4 May 1859 along with the Plymouth–Truro section of the railway. A goods shed was added in 1863, and the station was rebuilt in 1880; the surviving station building on the north side of the line dates from then. This building has been derelict for many years and although there were plans to redevelop it as a visitor centre the latest reports say that it has now been sold to a developer. The station canopy, goods shed and footbridge have gone and passengers using the Saltash station for local services are provided with a modern stone-built shelter on each platform. It is still well worth a visit as the platforms provide an excellent view of the bridge at rail level. Note how the two lines merge into one to cross the bridge.

Beyond Saltash the line turns due south and almost immediately comes to the site of the first of the Cornish viaducts over the tidal inlet at Coombe-by-Saltash. The original all-timber viaduct was of the type CLB4(D) with nine piers, the central five being timber trestles, supporting fans incorporating a queen post truss to the parapet level. The original viaduct was 603 feet in length, 86 feet high, and was built on a slight curve with a downward track inclination towards the southern end. In 1894 it was replaced by the present masonry viaduct and this can be seen when looking across the Tamar from the Devon side of the bridge.

Below: Map of Cornwall, *c.* 1900. Brunel's railways followed the southern route through Liskeard, via St Austell, Truro, Redruth, and Hayle to Penzance, plus a branch south to Falmouth. *(CMcC)*

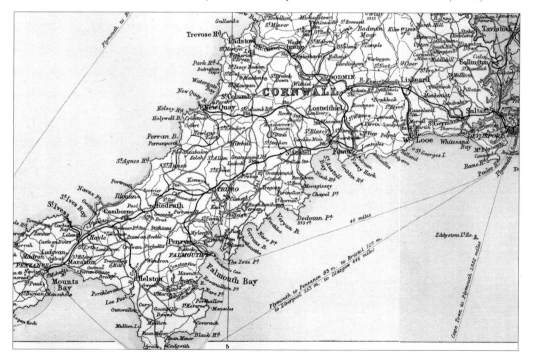

Above: Coombe viaduct, immediately south of Saltash, photographed from the car park on the Devon side of the Royal Albert Bridge. *Below:* The tranquil setting of the later Forder viaduct. The original had sixteen timber trestles and was replaced in 1908 when this part of the line was realigned.

South of Coombe the line begins a turn to the west above the St Germans or Lynher river, and after going through Shillington tunnel it crosses another tidal inlet at Forder Lake. This is a charming spot, but, as a local resident explained, 'the high tide comes in so fast here it is like a tsunami that washes up the shore'. The Forder viaduct was a QTTI(C) type, 606 feet long with sixteen slender timber trestles. The viaduct was not replaced until 1908 when the new line deviated slightly further inland with a masonry structure on a new alignment. The Forder Lake is accessed by a narrow single-track road. At one time the banks were full of orchards, and local legend says that the trains would stop by the viaduct and a basket was lowered on a rope to collect fruit for the market.

The line then crossed two more modest low-level viaducts, at Wivelscombe and Grove, and the 67-foot-high Nottar viaduct took it over the Lynhar River. All three were taken out of service by the 1908 realignment. The next major crossing spans the River Tiddy, a tributary of the Lynhar, at St Germans. This is a spectacular location and the present masonry structure gives some idea of the scale of Brunel's original. At 945 feet long the rails were carried aloft 106 feet above the river on fourteen composite timber trestles which rested on timber piles midstream with masonry plinths at each end above the high water mark. The main spans were 64 feet and 66 feet between the centre of the trestles and these were further strengthened with wrought-iron ties. Although very similar to the Coombe viaduct this was a non-standard design. The present masonry viaduct, located on the down side of the original, was completed in 1907 and is no less spectacular in this postcard-perfect setting.

St Germans station is of interest as Brunel's original station buildings in the Italianate style have survived. Today the main station building with stationmaster's office is privately

Below: Shillingham Tunnel, a little beyond Forder and before St Germans. *(CMcC)*

Above: The magnificent setting for the St Germans viaduct over the River Tiddy. The original timber structure consisted of fourteen trestles and was replaced by the present masonry viaduct in 1907.

Left: This GWR map dates from the 1920s and shows the line crossing the Tamar from Plymouth, and then swinging down and along the northern side of the River Lyner to St Germans. Brunel's original line had been slighty further south at this point and was realigned in 1908; it appears on the OS maps as a disused railway.

Above: St Germans still has its Brunel-era Italianate buildings, although the footbridge is a later addition. The station remains in use, but the booking hall and stationmaster's office are now a private dwelling.

Right: Just two of several old railway carriages in the yard at St Germans station. Photographed in 2013, they are being restored to create distinctive holiday accommodation. There is a fully converted example of a GWR travelling post office in the goods dock beside the western platform of the station.

Coldrennick or Menheniot viaduct is a tall structure visible from the A38 main road. As can be seen from these images, the timber elements were replaced by extensions to the masonry piers.

owned by David Stroud, and he uses the neighbouring yard to restore old railway coaches to create holiday accommodation. A completed example, a GWR travelling post office, stands in the goods dock beside the western platform. As it was customary on the early railways for passengers to cross the tracks on foot to get from one one side to the other, the footbridge at St Germans was not added until 1901. The station is still in use on the line operated by First Great Western.

From St Germans the route heads north-west towards Liskeard, roughly paralleled by the A38 road. At Tresulgan it originally crossed a tributary of the Seaton River on a 525-foot-long viaduct on eight piers, later replaced by the masonry structure in 1899. The next viaduct, Coldrennick, is a mile further on and is adjacent to Menheniot station. Visible from the main road the scale of this viaduct is immediately apparent, 790 feet long and 138 feet high. The original consisted of high masonry piers topped by fans of timber to support the rail deck in a CBL4(A) configuration. But unusually, when it came to a replacement, only the timber elements were removed and the existing masonry piers were raised with brick extensions to support latticed steel girders. In February 1897 this was the scene of a terrible accident when scaffolding used in the rebuilding collapsed, sending twelve workers plunging to their deaths. In 1933 the piers were encased in granite.

Below: Looking up at the decking of the Coldrennick viaduct, showing the extensions to the piers. In 1897 this was the scene of a terrible accident when the workers' scaffolding collapsed.

Menheniot station is immediately west of the Coldrennick viaduct. Unfortunately nothing remains of Brunel's main station buildings, shown above. These were demolished following a fire. However an original open-fronted waiting shelter has survived on the other platform.

Immediately to the west of the Goldrennick viaduct is Menheniot station, another original Brunellian station, but alas the main chalet-style building was demolished following a fire and only an open-fronted shelter survives on the far platform. A goods yard behind the station once served the neighbouring Clicker Tor Quarry, although the sidings were removed in the 1970s. The station is now a request stop for local trains.

Liskeard to St Austell

Before reaching Liskeard there are viaducts at Treviddo, Caruther and Bolitho, all of the CLB4(A) design and replaced by masonry ones in the 1980s and 1990s. The Liskeard viaduct crosses a deep valley immediately before the station itself. Built on a curve it was 720 feet long and 150 feet high, making it the second highest in Cornwall. It has eleven piers and at the time of its replacement, commencing in 1893 and completed the following year, these were built up with brickwork to support lattice steel girders beneath the rail deck. The track was doubled in 1896, and ironically the steel girders had to be replaced in 1929 owing to corrosion.

Liskeard station is another Brunel original, opened on the main line in May 1859. The station building is perched above the cutting, and in 2005 a refurbishment saw the addition of new facilities including an extended canopy and improved café. There's a lot going on at Liskeard including a fine post-Brunel signal box (*c.* 1915) and, at the southern end of the platform, and at right angles to the tracks, there is the smaller station for the Liskeard–Looe branch. But best of all for Brunel hunters is the flying-arch road bridge across the main line, springing from the sides of the cutting without abutments.

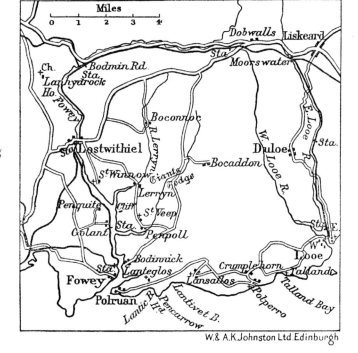

Right: Map of the route going westwards from Liskeard to Lostwithiel. The branch north-west from Bodmin Road goes to Wadebridge. The station at Bodmin Road opened in June 1859, with a short branch into Bodmin General, in the town, opened in 1887; although laid as standard gauge it did not become fully connected until conversion of the main line from broad gauge in 1892.

W.& A.K.Johnston Ltd.Edinburgh

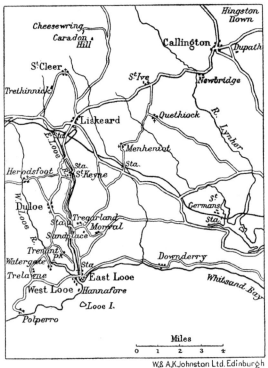

Above: Brunel's flying-arch road bridge at Liskeard, which springs from the side of the cutting without abutments. Liskeard is a fascinating station, and in addition to the bridge it retains its Brunel station building, a later GWR signal box, and a bonus second station at right angles to the main line for the Liskeard–Looe branch line.

Liskeard's viaduct is immediately south-east of the station and is one of the highest in Cornwall.

Left: Map showing the main line from St Germans up to Liskeard, and the branch looping from Liskeard station to head south to the coastal resort of Looe.

Continuing westward we encounter several fine viaducts. The first, at Moorswater, is visible from the A38 but tricky to get to on the minor roads that branch off it. Crossing the wide East Looe river, Moorswater has rightly been described as the most spectacular of all the Cornish viaducts. At 954 feet long and 147 feet high it leapt across the valley in a graceful curve, its fourteen crucifix-buttressed piers pierced by multiple pointed Gothic-style openings. Moorswater was built as a type CLB4(D) with double fans of timber and longitudinal ties supporting the tracks. Following concerns regarding the state of the structure, some of it in the form of letters published in local newspapers, Moorswater was one of the first major viaducts to be totally replaced. This work was completed in February 1881, barely twenty-two years after the opening of the railway. The new masonry structure is very elegant and consists of eight arches of 80-foot span. Beside it six of the original Brunellian piers still stand tall and proud, and the stumps of most of the others are still in situ, although No. 2 pier on the slope of the eastern side has toppled over.

Between Moorswater and St Austell the line passes over a number of viaducts, including several in quick succession:

Westwood, located in the woods on the valley side. 372 feet, 88 feet high, replaced by masonry viaduct in 1879. Access to this site is almost impossible, with quarry waste almost burying any remains of Brunel's piers.

St Pinnock viaduct, 633 feet long, 151 feet high, with nine buttressed piers. These were raised with pierced masonry extensions and new iron support girders in 1882. This fine viaduct is located behind the Trago shopping outlet on the south side of the A38.

Largin viaduct crosses the wooded Glynn Valley. 567 feet long, 130 feet high, with eight piers, it was replaced in a similar fashion to St Pinnock in 1886. The line has been reduced to a single track since 1964.

West Largin, 315 feet long, 75 feet high, with three piers and two fans coming from the slopes. Replaced by the masonry viaduct in 1875.

Draw Wood, 669 feet on a curve, 42 feet high, with seventeen stub piers. Replaced by masonry retaining wall and an embankment in 1875.

Below: The St Pinnock viaduct, which was raised in 1882 with new iron support girders.

Opposite and above: Built on a curve, the Moorswater viaduct looked magnificent. It was 145 feet high and 954 feet long, and had fourteen buttressed piers with Gothic-style openings. However, Moorswater may have been weak and in 1884 it became one of the first of the major viaducts to be replaced. Six of the original Brunel piers still stand beside the new, elegant masonry structure.

Above: Looking down upon St Austell station from the road. This view matches the old postcard shown opposite. It is a well-preserved station with a wooden building on the Up side, a later GWR signal box, and a rare surviving example of a covered GWR footbridge, shown left.

Left: A lesser-known Brunel landmark, the Grade II listed Carlyon railway bridge passes over the road to Carlyon Bay. It is located just to the east of St Austell.

Derrycombe, 369 feet, 77 feet high, with five piers. Replaced by masonry viaduct in 1881. Three of the original masonry piers are still standing.

Clinnick passes over a tributary of the Fowey. 330 feet long, 74 feet high, with five piers, it was replaced by the masonry viaduct in 1879.

Pendalake, the last of the Glynn valley group, is a low-level viaduct 426 feet long, 42 feet high, with ten sets of timber fans resting on low masonry plinths. It was replaced by a masonry structure in 1877.

Near Lostwithiel, a timber bridge consisting of a single span of just over 28 ft crossed the Fowley. This was followed by the Milltown viaduct, 501 feet long, 75 feet high, with five piers and two buttressed piers. Replaced in 1896 when the line was doubled.

St Austell to Truro

St Austell is described in the GWR's *Through the Window* guide – price one shilling – as the chief centre of the china clay industry. In Bradshaw's day the area was also known for the mining of granite, tin, copper and nickel. The town is blessed with a well-preserved station, albeit largely of the later GWR type, and a modern main building that was completed in 2000. The wooden building on the Up side has survived. At the time of its opening in May 1859 the station provided a passing loop on the otherwise single line. It was an important station for freight traffic and encompassed a goods shed, yards and assorted sidings by 1910, and then in 1931 an even larger goods yard was opened a mile to the east. St Austell has a rare example of an original GWR covered footbridge, festooned with the company's monogram. Note also the GWR type 7C signal box at the western end of the station, now disused and boarded up.

Above: Postcard view of St Austell station, *c.* 1905. The sidings on the right have given way to a car park area. The first of two viaducts is immediately beyond the station.

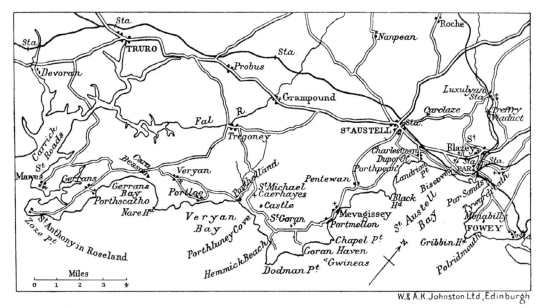

Above: GWR map showing the St Austell–Truro route.

Almost immediately to the west of the station there are two viaducts in quick succession, and both are VLB4(A) designs. The first is the St Austell viaduct, 720 feet long, 115 feet high, and built on a curve with ten masonry piers supporting timber fans originally. Located at the foot of the Trenance valley its scale is made all the more dramatic by the proximity of the houses that nestle at its feet. It was replaced by the masonry arched viaduct in 1898 and this runs beside the Brunel one for the most part, although due to lack of space the eastern, or Plymouth end, of the newer structure overlaps the old. Just a mile beyond St Austell the line crosses the Gover Valley. Also built on a curve, the Gover viaduct was 690 feet long, 95 feet high and featured ten masonry piers. It was replaced by a new masonry viaduct on the Down side in 1898.

Before reaching Truro the line is carried on another five viaducts. Coombe St Stephens was 738 feet long, 70 feet high and had eleven piers. Unusually it followed an 'S' curve and this may have resulted in structural weaknesses. Accordingly, it was replaced by a masonry viaduct on the Up side in 1886. The stumps of the original piers still remain.

Next is Fal viaduct, 570 feet long, 80 feet high, with eight piers. Another on an 'S' curve, it was replaced by the new viaduct in 1884 and only a lone Brunellian pier remains.

Probus, 570 feet long but relatively low at 43 feet high. It had eleven piers and was replaced by an embankment in 1871 making it one of the shortest-lived of the timber viaducts.

Tregarne, 606 feet long, 83 feet high, with seven piers plus a timber fan from each embankment. Replaced in 1902.

Tregagle was 315 feet long, 69 feet high and built on an incline of 1 in 66 with only four piers. It was replaced by a masonry viaduct on the Up side in 1902.

Opposite page: Two views of the St Austell viaduct, before and after replacement in 1898. In the recent photograph the original piers stand beside their replacement, nestling at its base. The proximity of the houses make it look all the more dramatic.

ST AUSTELL.

Truro

The cathedral city marks the meeting of the Cornwall Railway and the West Cornwall Railway and the two companies shared the station, which opened on 4 May 1859 following the completion of the Plymouth–Truro section of the Cornwall Railway. The following contemporary description comes from the *West Briton & Cornwall Advertiser*:

> The passenger station here is a handsome stone building, one hundred and thirty feet long, with large projecting roof; and containing in the centre of the building a spacious booking office, having separate entrances for first, second and third class passengers. On each side of this are comfortable first and second class waiting rooms, parcels' room, superintendent's office, and the other conveniences of a first class station. Inside the station is the passenger platform, one hundred and sixty-one feet long by fourteen feet wide, and beyond this three lines of broad gauge rails. Then the arrival platform, which is of the same length of that on the opposite side, and twenty feet wide. The whole of the space occupied by these rails and platforms are covered by a double roof, of the respective spans of fifty-seven and forty-one feet, with iron tie and suspension rods on a novel principle. The light, airy and forceful appearance of these roofs has excited the admiration of every person who has viewed them.

An extensive rebuild was completed in 1900 in a typically Victorian style. It features a hipped slate roof that seems entirely disproportionate to the remainder of the building, making it one of the ugliest stations in the whole of Cornwall. There have been numerous other changes to Truro station over the years, but at least it still retains the West Box dating from 1897 and located beside the level crossing.

Below: The late-Victorian station frontage at Truro, with a more recent extension.

10758. - TRURO, GENERAL VIEW.

Above: A view across the city with the long Truro (Moresk) viaduct clearly visible to the right of the cathedral. Truro's second viaduct, the Carvedras, on the left-hand side, is not so easily seen. *(LoC)*

While the station might not win any beauty awards, Truro more than makes up for its shortcoming as the undisputed viaduct city. It has two of the finest to be found, including the longest in the county. This is the Truro (Moresk) viaduct, 1,329 feet long, with a maximum height of 92 feet and incorporating twenty piers, of which six were buttressed. This is the mother of all CLB4(A) designs and forms a major landmark on contemporary photographs and postcards, striding across the city that already boasts a triple-spired cathedral. The old viaduct was replaced by a masonry one in 1904, and fourteen of the original masonry piers stand to attention beside it. You can get up close at the car park off Moresk Road and Oak Way. With only the briefest interval the line passes over Castle Rise before it is borne aloft on yet another viaduct, the Carvedras, which was 969 feet long, 86 feet high and had fifteen piers. Towering over Victoria Park and St George's Road, this is one of the most photographed of the timber viaducts and featured on numerous postcard views. Another CLB4(A) type, the piers comprised three heavily buttressed and eight plain while the landward spans rested on masonry abutments. The Carvedras viaduct was not replaced until 1902 and five of the original piers still tower over the gardens.

49

Above: Several of the original Brunellian piers stand beside the replacement Truro viaduct. It is surprising that the old piers were not incorporated within the newer structure, which was completed in 1904.

Opposite page: Two views of the Carvedras viaduct at Truro, looking south down St Georges Road. In the upper photograph it is possible to get a sense of how precarious a journey across this single-track viaduct must have felt, riding high above the rooftops. The timber viaducts were not popular with passengers, especially when they creaked under the load of a broad gauge train. Its replacement was completed in 1902, and five of the original piers remain.

Truro, Railway Viaduct

51

The Waterfall Bridge and Viaduct Truro.

Left: The accessibility of the Carvedras, with St Georges Road and Victoria Park at its feet, saw it featured on countless postcards.

Opposite page: One of the five surviving piers at Carvedras.

St. Georges Road & New Viaduct, Truro

Bottom left: Buckshead Tunnel is on the eastern side of Truro, shortly before the line is carried on the Truro (Moresk) viaduct. *(CMcC)*

VICTORIA GARDENS →

Left: Looking down the station approach road towards the replacement viaduct at Redruth. Crossing over the town, the original featured long timber supports fanning upwards from stub masonry piers.

Above and left: Looking south-west at Redruth station. Although the original main building has gone, it is still a very pleasing station with its GWR footbridge. The monogram on the bridge might be mistaken for the West Cornwall Railway, but the date 1888 confirms that this is definitely a 'G' and not a 'C'.

The West Cornwall Railway

The West Cornwall Railway was incorporated by Parliamentary Act in 1846 as a continuation of the Cornwall Railway from Truro to Penzance. The line incorporated the earlier Hayle Railway – an early mineral line opened in 1837 to serve the mines in the Redruth and Camborne areas – and extended it at both ends, but in order to save costs it was permitted to continue as a narrow or standard gauge line. In 1850 the Cornwall Railway obtained new powers under which it could give six months' notice to the WCR for a third rail to be laid for mixed or broad gauge running. This they duly did in 1864 and the Joint Committee of Associated Companies, consisting of the Great Western Railway, Bristol & Exeter Railway, Cornwall Railway and the South Devon Railway, agreed to take over the railway in order to cover the cost, allowing through traffic to run all the way from Paddington to Penzance for the first time in early 1867. The Associated Companies were amalgamated as the GWR in 1876 and the West Cornwall Railway became subsumed within the GWR network.

In comparison with the terrain traversed so far, the line to the east of Truro, on the final stretch down to Penzance, featured far less dramatic undulations on the whole, and consequently its viaducts, nine in total, were all relatively low level.

Truro to Penzance

The first viaduct beyond Truro is Penwithers, 372 feet long on a curve, with a height of 54 feet. It was completed in 1852, the timber supports replaced by masonry in 1869, and a replacement for double tracks built on the Down side in 1887. There are no remains of the original Brunel viaduct.

Chacewater viaduct was also on a curve, 297 feet long, 68 feet high with six timber piers. Built in 1852 it remained in service until 1888.

The Blackwater viaduct, 396 feet long and 68 feet high, had eight sets of timber supports sitting on short masonry piers. Completed in 1852, it was replaced in 1888. This is the only viaduct on the West Cornwall section that has the remains of one of the original Brunellian masonry piers. See Penzance viaduct for an explanation.

Next comes Redruth, and despite being mostly post-Brunellian, the railway station here is still a little gem. Located at the top of a hill rising from the base of the viaduct, it was opened by the WCR on 11 March 1852 and replaced an earlier Hayle Railway station on the west side of town. The present main station building was constructed by the GWR in the 1930s to a standard design with a wide canopy. It is thought that the wooden shelter on the westbound platform dates from 1888. There used to be a large wooden goods shed on the Down side, but this has gone and the site is now a car park.

Of Redruth itself Bradshaw informs us:

> This town derives nearly all of its importance from its central situation with respect to the neighbouring mines, the working of which has increased the population to treble its original number, as nearly all the commercial transactions of the miners are carried on here.

On the south-west end of the station the line leaps off on the Redruth viaduct, flying over the streets, with the 90-foot-high Basset Monument atop Carn Brae tor in the distance. At 489 feet long and 61 feet high, the viaduct was built in 1852, to be replaced by a masonry viaduct in 1888. This more or less followed the line of the original but it is possible to see some of the masonry stump piers set off slightly at the base of the present granite structure.

Above: The remains of one of the original stub piers at the foot of the newer masonry structure. The lower photograph shows the surviving timber waiting room on the station.

Opposite: The Redruth viaduct straddles a main road leading into the town. The viaduct is immediately south-west of the station, and the lower view from the end of the platform shows the Carn Brea monument in the distance. Note the sign for HST drivers.

HAYLE — _St Ives in the distance_

Above: This engraving of the original Hayle viaduct shows the simple nature of its construction. The tall chimneys are part of the Harveys engineering works. The replacement viaduct, consisting of a row of slab-sided piers, is shown below on a rainy summer's day in 2013.

Above: The long viaduct at Hayle runs across the top of Penpole Creek. *(John Bennett)*

Between Redruth and Hayle there are three further viaducts:

Built on an embankment of the Hayle Railway in 1852, Penponds viaduct was 693 feet long and 45 feet high. The replacement was erected on the Down side in 1900, making this the last of the timber viaducts of West Cornwall. Next comes Angarrack viaduct, which has been described as the most impressive of the West Cornwall viaducts. It was was 798 feet long, and at 100 feet high it towered over the village. Built in 1851 the viaduct was replaced in 1886. Guildford viaduct was 370 feet long and 56 feet high. It was originally built in 1850 with timber footings, although these were later changed to masonry stub piers in the 1860s. It was replaced by a granite viaduct in 1886.

The Hayle Railway had been taking tin and copper ore from the mines at Tresavean and Lanner to the foundry and wharves at Hayle since December 1837, and was incorporated into the route of the WCR in 1852. The long, curving timber viaduct at Hayle was constructed to carry the line across the town on its way to Penzance. Located just beyond Hayle station, 831 feet long and 34 feet high, its thirty-six spans were supported on simple two-leg timber supports for most of its length, with those in the immediate proximity of the Harveys engineering works and foundry at the Penzance end constructed in masonry to reduce the risk of fire. (Local street names such as Foundry Hill, Foundry Lane and also Foundry Square are reminders of the town's industrial past.) The timber viaduct was replaced on alignment in 1886 with masonry piers. These slab-sided piers look like a row of massive dominoes as they straddle the top end of Penpole Creek, and several have developed a slightly lopsided attitude over the years, with the line held level by packing beneath the support girders.

The original station serving the Hayle Railway had opened in May 1843 in Foundry Square, but was closed in 1852 to be replaced by the WCR station at the eastern end of the viaduct. Mixed gauge traffic arrived in 1866 with the addition of a third rail, and the track was doubled in 1892 following its removal. Although some of the WCR station buildings were retained into the GWR era, they, together with the goods shed and Hayle East and West signal boxes, have all gone.

South of the town the line passed over the River Hayle near St Erth via a simple timber double-span bridge of 85 feet, later replaced by a steel girder bridge supported on granite piers and abutments.

59

S^T MICHAEL'S MOUNT

Above: The scale of St Michael's Mount has been somewhat exaggerated for dramatic effect in this engraving, which also shows the railway running along the shoreline with a train out of Penzance.
Below: In December 1852 the timber viaduct was badly damaged by heavy storms.

Into Penzance

This long, low-level viaduct, with a maximum height of only 12 feet, carried the line along the shoreline of Mounts Bay into Penzance. Built in 1852, it was 1,041 feet long and consisted of fifty-one 19-foot spans. Standing on the edge of the beach, the exposed structure had absolutely no protection from heavy seas and on 27 December that same year a great storm caused considerable damage, as reported by *The Illustrated London News*:

> During the gale the whole of the scaffolding and wood-work erected for the works extending the southern arm of Penzance pier, were wholly swept away; and the scaffolding was drifted by the heavy sea with tremendous force against the wooden viaduct of the West Cornwall Railway, on the Eastern Green, just at the entrance of the town, and a considerable portion was destroyed. On that part of the line still nearer to the town, a considerable portion of the sea wall which there protected the railway was thrown down, and about 180 feet of the viaduct was swept away, or so far mutilated as to require replacing.

Until repairs were made the trains came only as far as Marazion station, and omnibuses conveyed passengers the rest of the way to Penzance. Winter storms severely damaged the viaduct once again, in 1868, and the line was diverted on temporary lines until a new wooden viaduct was completed in 1871. In 1921 this was finally replaced by a wider, granite-faced embankment to accommodate the doubling of the track, and it is likely that much of the stone used to build it came from the masonry piers of the redundant timber viaducts on the western section of the railway. This explains the absence of original masonry on these viaducts.

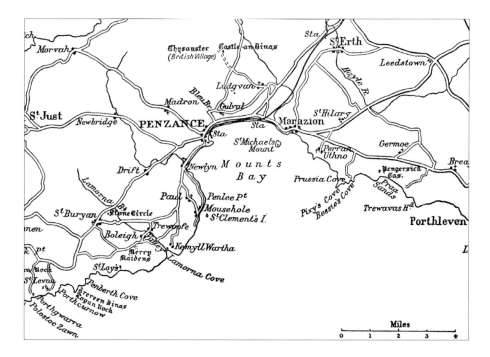

The end of the line.
Top left: The Illustrated London News engraving of the newly opened terminus at Penzance, published in 1852. The goods shed is on the left of the station.

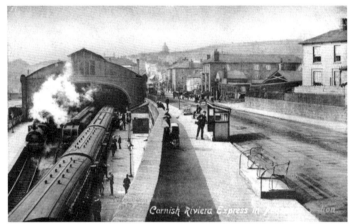

Left and below: Two postcard views of the enlarged station. Both feature the flagship 'Cornish Riviera Express'. In the lower image the corner of the timber goods shed is to the far left.

Above: A recent view of the station at Penzance. The all-over roof has been rebuilt and the goods shed demolished.

Right: St Michael's Mount on the cover of the popular GWR publication, '*Through the Window' Paddington to Penzance.*

63

Above: The timber viaducts on the Falmouth branch were the last to be replaced. This is the Carnon viaduct during the rebuild in 1933. The upper photograph shows the original structure with the new piers taking shape alongside.

Left: The reconstruction of the Carnon viaduct is underway.

The Falmouth Line

With the main line from Truro going via St Austell to Penzance, the 12-mile extension to Falmouth eventually became little more than a branch line. Consequently, while the main line opened in May 1859, it would be another four years before work on the Falmouth line was completed in August 1863. All of the viaducts in this section were opened with the line itself, and although Brunel had died in 1859, under the guidance of Robert Brereton they still follow his standardised CLB4(A) design produced for the Cornwall Railway.

The railway leaves the West Cornwall main line at Penwithers Junction, about half a mile to the south-west of Truro station, although as the intended main line it continues in a straight line while the Penzance line curves off to the right. At Penwithers viaduct work had commenced in 1853 with four of the piers partly built, and it didn't recommence until 1861. The original viaduct was 813 feet long, 90 feet high, and had ten piers with a landward fan sitting on buttressed plinths at either end. In 1926 it was finally replaced by a substantial embankment. The line then enters Sparnick tunnel, which is a little over a quarter of a mile long. Next came Ringwell viaduct, a relatively short one at 366 feet, and 70 feet high, with only three piers and a fan rising from the embankment slope at either end. As with Penwithers, it was also replaced by an embankment, slightly to the north, started in 1932 and completed the following year.

Right: The branch departs from the main line a little to the west of Truro, and heads southwards down to Falmouth. Today it is promoted as the Maritime Line.

Above: A recent photograph of the Carnon viaduct showing one of the surviving piers.

Left: A GWR train crosses the newly completed replacement Carnon viaduct.

Opposite page: The Cornwall Railway goods shed at Perranwell station.

The Carnon viaduct was an especially impressive structure as it spanned the wide valley above the heavily silted inlet of Restronguet Creek. The valley has a soft sandy bottom and it was necessary to sink cast-iron cylinders, 16 feet in diameter, down to the rock bed below. The sand was excavated from the cylinders and foundations built, and then the cylinders, built in several sections, could be removed. This operation was repeated to provide the wide footing for each pier. The viaduct was 756 feet long, 96 feet high, carried on eleven piers. In 1933 it was replaced by a 'handsome' masonry viaduct with nine arches. The original masonry piers still stand and there is good access with a parking area and a footpath which follows the line of the old Redruth & Chasewater Railway leading under the viaduct.

Next came the Perran viaduct. Located within a densely wooded valley, this was 339 feet long, 56 feet high, and had five piers. It was replaced by a masonry viaduct in 1927.

Sandwiched between the Perran and Ponsanoth viaducts is Perranwell station, which opened on on 24 August 1863 as Perran, but was renamed in February the following year to avoid confusion with nearby Penryn. The old Cornwall Railway goods shed still stands beside the forecourt, but otherwise the station is unremarkable as nothing remains of other buildings. Ponsanooth is the highest viaduct on the Falmouth Branch, and some work had commenced in 1854, although it was not completed until 1861. Ponsanooth viaduct, 645 feet long and 139 feet high, had nine piers and crossed the River Kennall in a deep valley to the east of Ponsanooth. It was replaced by the present masonry structure, which was built alongside in 1930. It is a difficult viaduct to get up close to as the view is obscured by the surrounding woodland.

Above: Modern shelter at
Perranwell station.

Left: Construction work
in progress on the new
viaduct at Ponsanooth on
the Falmouth branch line.

The construction of the brick arches for the new Ponsanooth viaduct, above, is shown nearing completion in 1930. The original was a relative newcomer, being built in 1863. Nowadays it is difficult to get a good view of the viaduct at Ponsanooth from within the heavily wooded valley.

Continuing south, the line crosses Pascoe viaduct. At 390 feet long and 70 feet high, this had six piers and remained in service until 1923. It was replaced by the embankment built alongside.

Penryn viaduct, overlooking Falmouth and the harbour, has featured on many postcards. It was 342 feet long, 83 feet high, and had five masonry piers. In 1923 this viaduct was also replaced by an embankment, realigned to the east side, which required Penryn's station to be rebuilt too. All that remain now are the platforms and waiting shelters.

Collegewood is the final viaduct before the line reaches Falmouth and it is also the longest on the Falmouth Branch at 954 feet. It was 100 feet high, had fourteen piers, and continued in service until 1934, making it the last of Brunel's timber viaducts to be replaced.

Falmouth

Falmouth Docks station, the terminus of the Truro–Falmouth line, was opened as Falmouth station in August 1863. Sekon informs us:

> The Falmouth station is nearly a mile beyond the town, quite a unique experience, but just as inconvenient as the usual one of building a station a mile before the town is reached. Its position, however adjoins the docks which were expected to become of great importance.

The main station building was built in granite in the Brunellian style with an overall wooden roof, dismantled in the 1950s. It was closed to passenger traffic in December 1970 when a new station, Falmouth Town, was opened nearer to the town, but since then it has been reopened as the terminus for passenger services on the branch, which is now known as the Maritime Line. Local trains continue to operate between Falmouth and Truro, with the latter continuing to provide a connection with the main line.

It is interesting to note that the conversion of the GWR's broad gauge to standard also meant that the Falmouth Docks Company also had to narrow its 5 miles of lines. Under the superintendence of two experienced foremen of the GWR the works at the docks was completed just a day after the main railway lines.

Right: Arrival of a broad gauge train to mark the opening of the Falmouth Docks station in August 1863.

Opposite page: Three views of the line into Falmouth, showing the Collegewood viaduct, top, and the Falmouth Docks station.

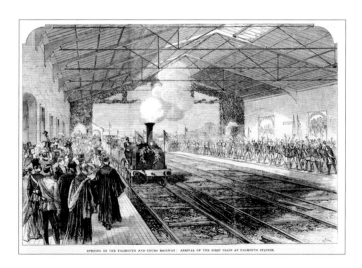

OPENING OF THE FALMOUTH AND TRURO RAILWAY: ARRIVAL OF THE FIRST TRAIN AT FALMOUTH STATION.

Above: Bristol & Exeter Railway broad gauge loco No. 44 was designed by the company's mechanical engineer, James Pearson, and built by Rothwell & Co. of Bolton in 1854. These massive 4-2-4T locos featured 9-foot driving wheels. *Below:* Replica of the GWR's *Iron Duke* at Toddington, Gloucestershire.

Cornwall's Broad Gauge

The railways of Cornwall have had an interesting relationship with Brunel's broad gauge. As we have seen, some parts of the line began life with what the more 'narrow' minded like to refer to as the standard gauge, but were broadened for a time only to be returned to standard at the broad gauge's demise, while most others were originally built as broad gauge.

At its height the broad gauge network stretched from London to Penzance, from the south coast to the Midlands and across southern Wales. In terms of scale this must make it Brunel's biggest and costliest failure. Yet it is hard to condemn a system that was actually superior to the one it was eventually replaced by. This wasn't a case of a good idea giving way to a better one, it was all about who had the greater weight of numbers on their side in terms of territory covered. When Brunel had been appointed as engineer to the Great Western Railway in March 1833, he determined that it would be the finest railway in Britain. At that time the railways were so new that there were no ground rules laid down by Government or any other body regarding engineering details such as the choice of gauge, the distance between the rails. In the north of England George Stephenson had chosen the 4 feet 8.5 inches for the Stockton & Darlington Railway for no other reason than that it was already used in the collieries which the line served, and naturally enough his son, Robert Stephenson, followed suit. That all seemed a long way from the Bristol–London railway that Brunel was designing. He dismissed the northern 'coal wagon' gauge in favour of a wider one at 7 feet and a quarter inch as being 'more commensurate with the mass and velocity to be attained'. One of his early proposals had been that the railway carriages would be contained within the wheels rather than sitting above them, although this was never put into practice.

It has been suggested that Brunel chose a different gauge for the sake of being different from his rivals, but whether or not this is true the 'broad gauge', as it became known, was

ratified by the GWR Board in 1835. The GWR would pioneer its own way and it was going to be magnificent. Brunel also had his own method of supporting the rails using longitudinal timber sleepers running the whole of their length, joined by cross sleepers at 15-foot intervals and anchored by beech piles driven into the ground. Once trains actually started running on these tracks it was discovered that the rigidity of the piling resulted in an extremely bumpy ride as the sections between them flexed up and down. The piles were pulled up or driven down clear of the track.

Over the next ten years, from 1835 to 1845, the various railway networks spread their spidery fingers across the map of Britain and the GWR extended westward beyond Bristol. Any new lines connecting to it had Brunel as their engineer and were also built to the broad gauge. It is sometimes overlooked that Brunel was not alone in adopting a 'non-standard' gauge. About a hundred miles of the London & North Eastern Railway had been constructed to a gauge of 5 feet, some Scottish lines had been laid with 4-foot-6-inch and 5-foot-6-inch gauges, and the Surrey Iron Railway had a gauge of 4 feet. Opinions varied on the merits of the various gauges, both in terms of performance and the desirability of standardisation, but matters only came to a head when different gauges met head-on. This happened for the first time at Gloucester where the Midland Railway met an extension of the GWR. Brunel's solution was the transit shed which had broad gauge track entering on one side of a central platform and the 'standard gauge', as it was already being referred to, on the other. All passengers, their luggage and goods had to be laboriously transferred from one train to the other. The resulting pandemonium was described by *The Illustrated London News*:

Above: The 'Flying Dutchman' express caught in the snow at Camborne in Cornwall. The saddle tank locomotive is *Leopard*. Camborne's most famous son is Richard Trevithick who built the world's first first railway locomotive in 1804.

Above: Utter confusion at a transfer station where the broad and standard gauges met head-on at Gloucester. An example of a transfer station can be seen at the Didcot Railway Centre.

> It was found at Gloucester that to trans-ship the contents of one wagon full of miscellaneous merchandise to another, from one Gauge to another, takes about an hour; with all the force of porters you can put to work on it ... In the hurry the bricks are miscounted, the slates chipped at the edges, the cheeses cracked, the ripe fruit and vegetables crushed and spoiled; the chairs, furniture, oil cakes, cast-iron pots, grates and ovens all more or less broken ... Whereas, if there had not been any interruption of gauge, the whole train would in all probability have been at its destination long before the transfer of the last article, and without any damage or delay.

The government called for a Royal Commission to look into the matter. Forty-four witnesses gave evidence on behalf of the standard gauge, while the broad gauge was defended by only four, including Brunel and Gooch. Trials between the two factions were conducted with trains travelling between Didcot and London, and the broad gauge came out on top. But with considerably more miles of standard gauge than broad, the Commissioners recommended that, for the sake of uniformity, 4 feet 8.5 inches should be the gauge for all public railways. In 1846 the Gauge Act dictated that all new lines had to conform to this standard gauge, unless they were extensions to the existing broad gauge network.

Although the use of mixed gauge lines kept the broad gauge trains working for another fifty years, it had many drawbacks, not least of which was the cost, and gradually the broad gauge lines were converted to standard. The conversion of the remaining London–Penzance line took place over the weekend of 21–22 May in 1892. On the Friday, 20 May, the final through train to Penzance, 'The Cornishman', departed from Platform 1 at Paddington at

Above: The historic scene as the final broad gauge 'Cornishman' express prepares to depart from Paddington station on Friday 20 May 1892. It was hauled by the *Great Western* locomotive, shown below in this colour postcard. The remaining broad gauge tracks on the London–Cornwall main line were converted over the weekend of 21/22 May.

10.15 a.m. It was hauled by the *Great Western*, a Rover class locomotive. All along the route the train's progress was marked by a volley of fog-signals laid on the tracks, and at each station crowds of admirers waited to catch their last glimpse of a broad gauge train. 'The Cornishman' was timed to arrive at Penzance at 8.20 p.m., but it was 9 p.m. by the time it reached its destination. (At Penzance a large crowd had taken the last local train to Truro to ride back on the last broad gauge passenger train.)

The last through passenger train going from Penzance to London was the mail train which had already departed earlier that afternoon. At Swindon the locomotive *Bulkeley* took over from the *Amazon* for the final leg of the journey, arriving at Paddington the following morning. The railway historian and devout broad gauge supporter, G. A. Sekon, described the scene in *A History of the Great Western Railway*:

> On arrival at Paddington the passengers on turning out, feeling that they had taken an indirect part in the final performance of the broad gauge service, lingered to see the emptied carriages back out of the station, in preparation for their last broad run to the sidings at Swindon.

The very final broad gauge train to depart from Penzance consisted of the rolling stock of 'The Cornishman' together with a breakdown van, and set off at 10 p.m on the Friday evening for Swindon. An inspector travelled on the train and at each station he ascertained that all broad gauge stock had been worked away and issued a notice to the stationmaster confirming that the last train from Penzance had left their station and that the engineering department could take possession of the line. The remaining track conversion was carried out through an extraordinary feat of organisation and was completed in thirty-one hours. A lasting image to emerge from this transition is the scene of a gang of workmen lifting the broad gauge lines at Saltash with the Royal Albert Bridge in the background.

The end of Brunel's big idea was marked by *Punch* which published a cartoon depicting the ghost of Brunel passing among the navvies as they carried out their work. Entitled 'The Burial of the Broad Gauge' it bade farewell with the words, 'Good-bye, poor old Broad Gauge, God bless you.'

Explore the broad gauge

• Apart from the structures and buildings already ready described, there are several replica broad gauge locos: *Iron Duke* is displayed at the Gloucestershire Warwickshire Railway in Toddington, Gloucestershire – www.gswr.com. At the Swindon GWR Steam Museum there is the *North Star* replica incorporating driving wheels from the original – www.steam-museum.org.uk. The GWR Society's Didcot railway centre has a working replica of Daniel Gooch's *Fire Fly*, plus a short section of mixed gauge track and a transfer shed – wwwdidcotrailwaycentre.org.uk.

Above: A gang of workmen converting the track at Saltash station with the familiar sight of the Royal Albert Bridge in the background. *Below:* Broad gauge graveyard at Swindon.

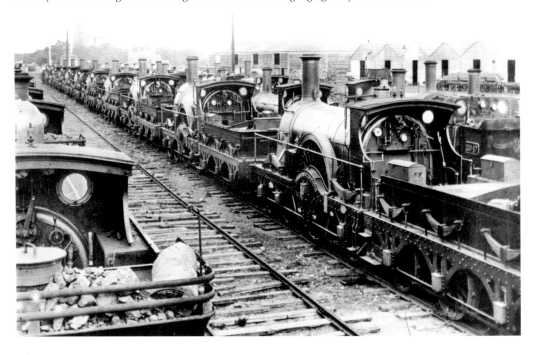

Punch's 1892 tribute to the passing of the broad gauge portrays the ghost of Brunel passing by as the workmen pay their respects to the end of an era.

Above: Brunel's second ship was the revolutionary iron-hulled, screw-driven *Great Britain*, launched in Bristol on 19 June 1843. His final great ship was the colossal *Great Eastern*, shown below. Both had Cornish connections.

Nautical Connections

Although none of Brunel's three great steamships – the *Great Western*, *Great Britain* and *Great Eastern* – ever put ashore in Cornwall, there are, nonetheless, a number of interesting Cornish connections.

On 12 June 1845 the *Great Britain* departed from Blackwall in East London, where she had been fitted out, on the first leg of a cruise around the coast to her new home port of Liverpool. The iron-hulled ship had been built in Bristol and had been launched there in June 1843. Following on from the wooden paddle steamer *Great Western*, also built in Bristol and launched there ten years to the day earlier, she was to continue the transatlantic service. It had been Brunel's intention that this would provide a connection for passengers travelling via the GWR to Bristol where they would board the ship direct to New York. Even though a hotel was built by the Great Western Steamship Company (GWSC) to accommodate them, and it still stands, the essential improvements to the entrance locks were not made in sufficient time for either side to be based in the city. The Great Western, when completed, had passed out of Bristol's Cumberland Basin into the River Avon with her paddles carried on her deck. The *Great Britain* had actually become stuck in the harbour entrance during the first attempt to leave in December 1844, and consequently both ships were operated out of Liverpool.

On that delivery voyage in June 1845, the *Great Britain* stopped briefly at Cowes on the Isle of Wight and then Plymouth which, incidentally is where Brunel's mother Sophia had grown up. The ship was greeted at Plymouth by a cacophony of church bells and the Hoe was crowded with thousands of onlookers. Further sea trials took place including a day trip around the Eddystone Lighthouse with 600 passengers on board. The Eddystone Rocks might be Devonshire territory, strictly speaking, but they are only 9 miles south of Rame Head, which is in Cornwall. The ship continued on to Liverpool, stopping at Dublin on her way, and commenced her first transatlantic voyage on 26 July. Although enthusiastically received by the New Yorkers, she was sailing with only a fraction of the cabins occupied, due to concerns about the safety of her iron hull and its possible effect on the ship's compass. After a winter refit to reduce her tendency to roll, the *Great Britain* made her third crossing to New York in May 1846. Passenger numbers were slowly rising but on the fifth voyage, which started from Liverpool on 22 September that year, disaster struck when the ship ran aground at Dundrum Bay on the Irish coast in County Down. Captain Hoskin claimed he had been confused by a newly commissioned lighthouse on the southern tip of the Isle of Man, but concerns remained that the compass was to blame. After surviving the wintry weather at Dundrum the *Great Britain* was refloated, but the incident had ruined the GWSC and both ships were sold off. The *Great Western* went to the West India Steam Packet Company, while the *Great Britain* was eventually sold in 1850, for a fraction of her original cost, to Gibbs Bright & Co. of Liverpool and she was refitted for the 12,000-mile journey to Australia where the discovery gold was attracting immigrants by the thousands. Converted primarily to a sailing ship with the passenger accommodation increased to 700, she made thirty-two voyages to Australia between 1852 and 1875. And it is on these voyages that we once again encounter Cornish miners.

On 12 June 1854 the *Great Britain* set sail for Australia with a party of 250 Cornishmen travelling in the cramped Third Class cabins, mostly tin miners drawn by the lure of gold. Conditions on the voyage through the Tropics were far from comfortable, with the ship's iron hull absorbing the heat of the sun, but the miners passed the time drinking, gambling, fighting and 'singing lusty hymns at Methodists services', according to Nicholas Fogg in *The Voyages of the Great Britain*. Inevitably mealtimes became a focal point of the day and tempers flared when the Cornishmen complained about the poor quality of the food. Some officers were threatened with knives, while one man took the plate he had been served and flung it into the face of Captain John Gray. On another occasion a large group started a drunken brawl in the bar. Captain Gray was knocked down and the First Officer is said to have drawn his sword.

No sooner had the situation on board calmed down than smallpox broke out. This was a disastrous turn of events as the Australian colonists imposed strict quarantine laws and the ship was ordered to anchor off Port Philip Head, some 40 miles out of Melbourne. An isolation station was established on shore and the ship was fumigated. Afterwards the First Class passengers were allowed to return to the ship, but the others had to remain in tents fashioned from a canvas sail until proper tents were sent from Melbourne. All passengers were inoculated and after almost twenty days their quarantine came to an end and the ship entered Melbourne to a cannonade of celebration that lasted several hours. On her return to England, the *Great Britain* was signed over to the Admiralty for conversion to a carry up to 1,650 troops to the Crimean.

Above: The *Great Eastern* was built at Millwall on the Isle of Dogs, London. These hefty timbers from one of the launchways have been uncovered at the site of the shipyard.

Above: Two views of the *Great Eastern.* Robert Howlett's photograph shows one of the checking drums holding the massive chains. The engraving from *The Illustrated London News* has the ship on its supporting cradles, refusing to enter the water. It took a Cornishman to eventually budge it.

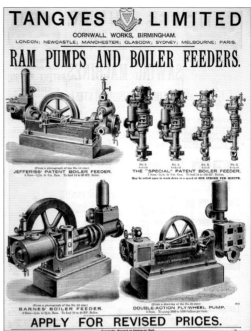

Richard Tangye was from Redruth, and together with his brothers he formed an engineering company in 1857 to supply equipment for the mines. It was their hydraulic jacks that pushed the *Great Eastern* into the Thames, inch by painful inch. One of the jacks is shown at Millwall, below.

Launching the *Great Eastern*

Brunel's third and final steamship, the *Great Eastern* was to be the largest ship afloat, a record that would last for almost half a century. Her vital statistics were impressive: 692 feet long, more than twice that of the *Great Britain*, and almost three times longer than the *Great Western*. Work began on the ship in the spring of 1854 beside the Thames at Millwall, on the Isle of Dogs, where she grew piece by piece until she loomed like a cliff-face of iron above the surrounding ramshackle buildings. In designing such a big ship IKB had put great emphasis on its strength. This was achieved with a double-skin, or rather double-bottom, hull with the iron plates set 2 feet 10 inches apart and containing longitudinal members spaced at 6-foot intervals. The expected method of construction would have been within a dry dock, which would have been flooded once the hull was completed, but the expense of creating such a big dry dock was considered too great. Instead Brunel opted for a sloping slipway constructed of massive oak timbers resting on piles. The ship's hull would sit in a cradle and this would slide sideways down the timbers into the river, held in check by chain drums. Similar launches had been tried before in the USA, but never on this scale and always as a free or unrestricted launch.

The day for the launch was set as 3 November 1857, a date dictated by the tides. Brunel conducted the proceedings with a system of signals, and when the time came for the steam winches to start there arose a thunderous sound as the chains reverberated against the hollow hull. The cheers of the crowd were suddenly cut short when the bow suddenly jerked forward, causing a wooden lever on one of the checking drums to flail out of control, killing one of the workmen. The launch was halted. On the next attempt on 19 November the ship did move, but only by about 14 feet. Only brute force could overcome the friction of the cradle against the wooden slipway. Powerful hydraulic jacks, or rams, were obtained from James Tangye & Bros, a little-known Birmingham-based engineering business which had been established in early 1857 by the Cornish brothers Richard, James and Joseph Tangye, from Redruth. Their initial customers had been the tin mines in the Redruth area. Twenty-one Tangye jacks were brought to Millwall to shove the ship, inch by painful inch, into the water and she was finally afloat on the Thames on 31 January 1858. Richard Tangye would claim afterwards:

> We launched the *Great Eastern* and she launched us.

In 1862 the Tangye company built its appropriately named Cornwall Works in Birmingham where it manufactured hydraulic jacks as well as steam engines, and later on produced industrial diesel engines, pumps and hydraulic equipment.

Cables

The *Great Eastern* had missed the boat for her intended role on the Australian run, and starting in 1860 she made only nine transatlantic round trips before being put up for sale by the shipping company, which had been making heavy losses. She was eventually bought for a mere £20,000 by a group of businessmen including Brunel's close friend and colleague Daniel Gooch. They set up the Great Eastern Steamship Company and the ship was refitted for a new role as a transatlantic telegraphic cable layer and chartered to the Telegraph

Left: The *Great Eastern*
was a failure as an ocean
liner. It was simply too
big for its time. However,
its size made it the ideal
vessel to lay a cable
across the Atlantic ocean.
The first attempt failed
when the cable broke,
shown left. A second
attempt, in 1865, was
more successful, and
Europe and the USA
were connected.

Left: The Telegraph
Museum at Porthcurno,
near Lands End, has
several *Great Eastern*
items on display
including this section of
the transatlantic cable.
Porthcurno became the
centre for international
telegraphic cables coming
into the UK.

Construction & Maintenance Company. On the first attempt the cable snapped mid-Atlantic, but on the second attempt in 1865 it was successful, and over the subsequent years the ship went on to lay cables from Brest, in France, to Saint Pierre and Miquelon, and also between Aden and Bombay. The main station for international submarine cables for Great Britain was established at Porthcurno near Lands End, a location chosen in preference to Falmouth because of the risk of ships in the busy harbour damaging the cables. The Eastern Telegraph Company was formed and built a cable office in the Porthcurno valley. The operation was expanded and by the interwar years fourteen cables were being operated simultaneously. In 1928 the ETC merged with the Marconi Wireless Telegraph Company Ltd to form Imperial & International Communications Ltd, which was renamed Cable & Wireless in 1934. The cable office closed in 1970, but a college established on the site remained until 1993. The buildings now house the excellent Porthcurno Telegraph Museum. Although the *Great Eastern*'s cable had terminated at the Telegraph Field on Valentia Island off the Irish coast, the ship's role in laying the first transatlantic cable is acknowledged in a number of exhibits at the Porthcurno museum and these include a piece of the cable.

At the conclusion of her cable-laying career the *Great Eastern* served as a show boat for several years, as a floating concert hall and gymnasium, and as a giant advertising hoarding. Sold as scrap in 1888, she was beached at Rock Ferry on the Mersey and broken up.

Below: The telegraph office at Penzance. Telegraphs were the most instant form of communication available to the Victorians, and remained so well into the twentieth century until the widespread arrival of the telephone. *(CMcC)*

Above: The *Great Britain* returns to Bristol in 1970. It was the only time that the ship would pass under Brunel's Clifton Suspension Bridge as it hadn't been completed when she left the city. *Below:* The *Great Eastern* spent her final days as a floating advertising hoarding. Here she is berthed at the North Wall in Dublin, plastered with faded advertising slogans. *(NLI)*

Above: At low tide the timbers of the *Great Eastern*'s launchway can still be seen in the Thames, beside the shoreline at Millwall on the Isle of Dogs.

Explore the *Great Britain* and *Great Eastern*

• In 1970, the *Great Britain* was returned to the Great Western Dock in Bristol, where she was originally built. You can explore the ship and the dockside buildings, which house a museum with many artefacts from Brunel's ships including both the *Great Britain* and *Great Eastern* – www.ssgreatbritain.org.

• The *Great Eastern*'s timber slipway can be seen at Millwall on the Isle of Dogs. Take the Docklands Light Railway to Mudchute or Island Gardens, and it is a short walk along the riverside. By car take the Westferry Road. Parking can be tricky on weekdays. You can also take the TfL London River Services boat to the Masthouse Terrace pier. Assorted relics and models are to be found at the National Maritime Museum in Greenwich, the Science Museum, London, and at the SS *Great Britain* in Bristol. One of the ship's masts has survived as a flagpole at Liverpool's Anfield stadium.

• The Porthcurno Telegraph Museum near Lands End has an excellent collection of telegraph equipment including examples of cables and a number of *Great Eastern* items – www.porthcurno.org.uk.

Brunel's Legacy

At first sight it might appear strange to talk of Brunel's legacy in Cornwall. The duchy is full of ghosts. The rows of discarded masonry piers that once supported the airy timber viaducts now stand like tombstones, and even the broad gauge, that great Brunellian experiment, is only a distant memory more than a century after it was unceremoniously ripped up. But that is to underestimate the significance of Brunel's imprint on Cornwall. We still have the Royal Albert Bridge, see below, and the main line from Plymouth to Penzance, not forgetting the branch lines too. Yes the timber viaducts have gone, but they were built to be disposable structures, and whether replaced with new timbers or by complete structures Brunel would have understood. As an engineer he was a pragmatist and realised that new needs and situations required new solutions. Engineering, and engineering works, continue to evolve and it is astonishing that so much of his work has survived at all or, at the very least, provided the foundations for today's railways. Cornwall is a rich hunting ground for Brunel enthusiasts and there remains so much to see to this day, not least the surviving and adapted viaducts, the many stations and of course his crowning achievement, the Royal Albert Bridge. The originals are all the more remarkable when you consider that they were built in a time before motor transport, before the railways, naturally, and before the telephones and computers that rule our modern lives.

Of all of Brunel's works, not just those within Cornwall but nationally, my favourite will always be the Royal Albert Bridge at Saltash. This jewel in his engineering crown remains little changed to this day. It is in constant use, and even the sleek HST 125 expresses are

Below: Great Western Railway map from the 1920s, showing the company's lines within Cornwall.

91

humbled by a 15 mph speed limit while crossing. The broad gauge track was removed from the bridge in 1892 and there have been a few minor modifications along the way including the addition of horizontal girders to strengthen the main spans, and the iron approaches being replaced by steel. More recently a £10-million refurbishment of the bridge has been completed. Work had started in 2010 and the engineers spent nearly 2 million hours of work strengthening and repainting the bridge. The latter process threw up an interesting discovery, and by stripping away countless layers of paint the engineers revealed that when completed in 1859 the spans had been finished with a pale stone or off-white colour. That didn't last long though, and red-brown anti-rust paint was slapped on within a decade, along with the 'goose grey' which we are more familiar with today. Following the refurbishment, the bridge is in great shape and it serves as a fitting tribute to the genius of IKB at the point of entry into Cornwall.

Looking beyond the physical remains, let's not forget the impact that the railways have had on the shape and nature of Cornwall's towns and cities, in particular with the coming of the tourism industry that is so important to its modern economy. Hopefully some of the visitors to the Cornish Riviera will also take a little time to explore the works of Victorian Britain's greatest engineer.

Sometimes it is the simplest tributes that say the most, and in conclusion I will return to the inscription applied to the portal of the Royal Albert Bridge by the directors of the Cornish Railway after his death:

'I. K. BRUNEL ENGINEER 1859.'

Opposite page: The portal of the Royal Albert Bridge on the Devon side, photographed in 2013 with the restoration work in progress. The section behind the portal is wearing its new paintwork.

The CORNISH RIVIERA

Further Reading

The Lost Works of Isambard Kingdom Brunel, John Christopher (Amberley Publishing, 2011).

Bradshaw's Guide – Brunel's Railways Paddington to Penzance, John Christopher (Amberley Publishing, 2013).

Brunel's Timber Bridges and Viaducts, Brian Lewis (Ian Allan, 2007).

Brunel's Cornish Viaducts, John Binding (HMRS Atlantic, 1993).

Brunel – In Love with the Impossible, Edited by Andrew and Melanie Kelly (Brunel 200, 2006).

Brunel – The Man Who Built the World, Steven Brindle (Widenfeld & Nicholson, 2005).

Isambard Kingdon Brunel – Recent Works, Edited by Eric Kentley, Angie Hudson and James Peto (Design Museum 2000).

The Life of Isambard Kingdom Brunel – Civil Engineer, Isambard Kingdom Brunel (Longmans Green, 1870).

The Voyages of the Great Britain, Nicholas Fogg (Chatham Publishing, 2002).

Cornwall Railway Stations, Mike Oakley (Dovecote Press, 2009).

A History of the Great Western Railway, G. A. Sekon (Digby Long, 1895).

Acknowledgements

I would like to thank the following individuals and organisations for providing photographs and other images for this book: Campbell McCutcheon *(CMcC)*, the US Library of Congress *(LoC)*, Network Rail, the National Library of Ireland *(NLI)* and John Bennett. Unless otherwise stated all new photography is by the author. *JC*.

Also in this series from Amberley Publishing

Brunel in Gloucestershire   ISBN 978 1 4456 0781 8
Brunel in Bristol   ISBN 978 1 4456 1885 2
Brunel in London   ISBN 978 1 4456 1885 5